Also by James Perloff:

The Shadows of Power: The Council on Foreign Relations and the American Decline (1988)

Tornado in a Junkyard: The Relentless Myth of Darwinism (1999)

THE CASE AGAINST DARWIN

WHY THE EVIDENCE SHOULD BE EXAMINED

James Perloff

BURLINGTON, MASSACHUSETTS

First printing, December 2002
Second printing, March 2003

Printed and bound in the United States of America.

Published by Refuge Books, 25 South Bedford Street, Burlington MA 01803

Much of the material in this book appeared in the July 2001 edition of *Whistleblower* magazine. Subscriptions to *Whistleblower* are available by calling (541) 597-1776 or online at www.worldnetdaily.com.

Book design, typography and electronic pagination by Arrow Graphics, Inc., Cambridge, MA

Cover design by Cameron Bennett and the Ingbretson Studio, Manchester, NH

Library of Congress Catalog Control Number: 2002092580

Cataloging-in-Publication Data

Perloff, James
The Case Against Darwin/James Perloff—1st ed.
p. cm.
ISBN 0-9668160-1-3
1. Evolution. 2. Creationism. I. Title.
QH366.2.P47 2002
576.8—dc21 2002092580

Contents

CHAPTER 1

Is Darwin's Theory Relevant to Our Lives?

"What? 'The case against Darwin'? Are you nuts? Everyone knows evolution is a proven fact—just like the laws of gravity!" That's how some people respond when you tell them that there is a strong case today against Darwin's theory of evolution. In fact, that response is very understandable. It's how I used to react myself. After all, the theory has been taught for so many years in schools and colleges that many take its truth for granted.

However, in recent years, growing scientific evidence has arisen challenging the theory. That's what this book will discuss—in easy-to-understand language.

But before getting to that, we should probably first address another matter: Why should the average person care about evolution in the first place? Doesn't it just involve a lot of "science and biology stuff"?

One reason Darwinism is important to know about is its unique *social* consequences. Until the nineteenth century, the view almost universally accepted in the West was that God had created the world and man. Society, in turn, was largely structured on values laid out in the Bible.

Darwin's theory said something dramatically different: that man was not created by God, but evolved from ape-like ancestors, and that life itself was not created by God, but happened because the right chemicals came together by chance in the ancient ocean. After publication of *The Origin of Species* in 1859, Darwinian ideas began replacing religious ideas until evolution itself became the prevailing view.

Now it is certainly true that some people tried to make their religious beliefs compatible with evolution—they would say something like, "Well, maybe God created life through chance and evolution." However, for many, Darwin's theory put God out of the picture—he was irrelevant, or didn't exist. Julian Huxley, one of evolution's foremost spokesmen in the twentieth century, stated that "Darwinism removed the whole idea of God as the Creator of organisms from the sphere of rational discussion."[1] And that's how many people saw it. In fact, evolution has produced a lot of atheists—including some famous ones.

Soviet dictator Joseph Stalin murdered millions. In 1940, a book was published in Moscow entitled *Landmarks in the Life of Stalin*. It states:

> At a very early age, while still a pupil in the ecclesiastical school, Comrade Stalin developed a critical mind and revolutionary sentiments. He began to read Darwin and became an atheist.
>
> G. Glurdjidze, a boyhood friend of Stalin's, relates:
>
> "I began to speak of God. Joseph heard me out, and after a moment's silence, said:
>
> "'You know, they are fooling us, there is no God. . . .'
>
> "I was astonished at these words. I had never heard anything like it before.

"'How can you say such things, Soso?' I exclaimed.

"'I'll lend you a book to read; it will show you that the world and all living things are quite different from what you imagine, and all this talk about God is sheer nonsense,' Joseph said.

"'What book is that?' I enquired.

"'Darwin. You must read it,' Joseph impressed on me."[2]

Karl Marx said: "Darwin's book is very important and serves me as a basis in natural science for the class struggle in history."[3] Marx even sent Darwin proof-sheets of *Das Kapital*, and offered to dedicate it to him, but the naturalist politely declined, noting it might embarrass some members of his family.

Now Marxists may object to picking on them, so let's switch to arch-capitalism for a moment. Andrew Carnegie, the steel magnate, was once said to be America's richest man. Though raised a Christian, he became an atheist. There is a story that Carnegie returned to his native Scotland and was boasting to a crowd of poor folks about his great wealth. "Name one thing for me," he said, "that God could have given me that I haven't been able to get for myself!" An old man near the back of the crowd answered: "Well, I'll tell you one thing he could have given you, Mr. Carnegie—a sense of humility."

How did Carnegie become an atheist? He wrote in his autobiography:

> When I, along with three or four of my boon companions, was in this stage of doubt about theology, including the supernatural element, and indeed the whole scheme of salvation through vicarious atonement and all the fabric built upon it, I came fortunately upon Darwin's and Spencer's works. . . . I remember that light came as in a flood and all was clear. Not only had I got rid of theology and the supernatural, but I had found the truth of evolution.[4]

One person almost universally denounced is Adolf Hitler. While Marx saw the "struggle for existence" as between classes, Hitler saw it as between races, and sought to develop a "master race." But did he invent the idea? The subtitle of *The Origin of Species* was *The Preservation of Favoured Races in the Struggle for Life*. Although

Darwin penned that in an animal context, extending it to human races was a small leap of logic. In his demented way, Hitler was fulfilling this prediction Darwin made in *The Descent of Man*:

> At some future period, not very distant as measured by centuries, the civilized races will almost certainly exterminate, and replace, the savage races throughout the world. . . . The break between man and his nearest allies will then be wider, for it will intervene between man in a more civilized state, as we may hope, even than the Caucasian, and some ape as low as a baboon, instead of as now between the negro or Australian [aborigine] and the gorilla.[5]

Racism was very prevalent among leading early evolutionists, many of whom believed the races had evolved separately. Britain's Thomas Huxley, whose fierce advocacy of evolution won him the nickname "Darwin's bulldog," wrote:

> It may be quite true that some negroes are better than some white men; but no rational man, cognizant of the facts, believes that the average Negro is the equal, still less the superior, of the white man. And if this be true, it is

> simply incredible that, when all his disabilities are removed, and our prognathous relative has a fair field and no favour, as well as no oppressor, he will be able to compete successfully with his bigger-brained and smaller-jawed rival, in a contest which is to be carried out by thoughts and not by bites. The highest places within the hierarchy of civilization will assuredly not be within the reach of our dusky cousins. . . .[6]

Darwin wrote in *The Descent of Man* that "the weak members of civilized societies propagate their kind. No one who has attended to the breeding of domestic animals will doubt that this must be highly injurious to the race of man. . . . excepting in the case of man himself, hardly anyone is so ignorant as to allow his worst animals to breed."[7] Darwin's son Leonard became president of Britain's Eugenics Education Society—eugenics, of course, was the campaign to transform humanity through selective breeding. The German philosopher Friedrich Nietzsche, who advanced the idea of the "superman" and master race, called Darwin one of the three greatest men of his century. Zoologist Ernst Haeckel, probably Darwinism's greatest popularizer in Germany, wrote in 1904:

> The mental life of savages rises little above that of the higher mammals, especially the apes, with which they are genealogically connected. . . . Their intelligence moves within the narrowest bounds, and one can no more (or no less) speak of their reason than of that of the more intelligent animals. . . . These lower races (such as the Veddahs or Australian negroes) are psychologically nearer to the mammals (apes or dogs) than to civilized Europeans; we must, therefore, assign a totally different value to their lives.[8]

Thus Hitler did not invent his deadly racism—these ideas were simmering in Germany during his youth, and easily trace to Darwinian roots. As German philosopher Erich Fromm observed: "If Hitler believed in anything at all, then it was in the laws of evolution which justified and sanctified his actions and especially his cruelties."[9] Sir Arthur Keith, president of the British Association for the Advancement of Science, wrote in the 1940s: "The German Fuhrer, as I have consistently maintained, is an evolutionist; he has consciously sought to make the practice of Germany conform to the theory of evolution."[10]

This is not to in *any way* imply that today's evolutionists are racists; and certainly, Hitler's atrocities would have revolted Charles Darwin. But it *is* to say Darwinism has had relevant social impact. If people are only animals, then for Stalin and Hitler it made sense to *treat* them like animals, herding them like cattle into boxcars bound for gulags and concentration camps.

Now some may say: "Ah, well, Stalin, Marx, Carnegie, Hitler—those are just a bunch of old dead guys. What's that got to do with anything today?" It is true that we haven't had any Stalins or Hitlers running America. The U.S. Constitution decentralizes power, making it very difficult to form that kind of dictatorship.

But it is probably reasonable to at least say that we have experienced a *moral decline* in America over the last few decades. Not everyone would agree with that. Some people don't think morality can be defined. And certainly we have had recent *improvements* in this country—in, say, technology and some civil rights. But if we look at statistics such as drug use, teen suicide, and divorce, we see indications that the USA is declining. What happened at Columbine High School would have been unthinkable in the 1950s, when nobody dreamed that weapons detectors would ever be needed at school entrances.

So what's caused America's moral decline? Many would say, "Well, we've lost our respect for traditional values." OK, where did "traditional values" come from? They came mostly from the Bible, which for centuries was Western culture's central guiding document.

So we might more accurately ask: Why have we lost respect for the Bible? It is probably not an exaggeration to say that, above anything else, it was the widespread teaching of Darwin's theory of evolution as "fact." As Huxley said, evolution removed God "from the sphere of rational discussion." Once you make God irrelevant, the Bible becomes irrelevant, and the moral values *in* the Bible become irrelevant.

Our purpose in making this point is not to "push the Bible on people," but only to note that religion traditionally played a strong role in American social life, and that evolution tended to negate that role, with powerful effects of its own.

I used to wonder why America had such a big social transformation back in the sixties. One factor: Evolution was not heavily underscored in American public schools before then. But in 1959, the 100th anniversary of the publication of *The Origin of Species*, the National Science Foundation, a government agency, granted $7 million to the Biological Sciences Curriculum Study, which began producing high school biology textbooks with a strong evolutionary slant. In the meantime, the Supreme Court ruled that school prayer was unconstitutional (after having been *constitutional* for more than a century and a half). From then on, students in public schools heard the evolutionist viewpoint—man is just an animal—almost exclusively.

I wasn't raised religiously myself, but once sold on the "fact" of evolution, faith stood no chance with me—I became a hardcore atheist. And there was a reason why my generation, the baby boomers, accepted evolution so easily. Teenagers usually aren't too hot about morality to begin with. But here was *teacher* saying the Bible was an old myth. Well, to us that meant the *Ten Commandments* were a myth. We could make up our own rules! For rebellious teens, the message wasn't hard to take.

Harvard professor E. O. Wilson writes: "As were many persons from Alabama, I was a born-again Christian. When I was fifteen, I entered the Southern Baptist Church with great fervor and interest in the fundamentalist religion; I left at seventeen when I got to the University of Alabama and heard about evolutionary theory."[11] That's a pretty good summary of what happened to the baby boom generation.

Will Durant, author of *The Story of Civilization*, was one of the preeminent historians of our time. Shortly before his death, he said: "By offering evolution in place of God as a cause of history, Darwin removed the theological basis of the moral code of Christendom. And the moral code that has no fear of God is very shaky. That's the condition we are in."[12]

I know that not everyone will agree with the conclusion, but I hope we've made a reasonable case that teaching Darwin's theory as a fact has had some serious social consequences. "But surely," someone may ask, "you don't advocate attacking the theory of evolution just because of *that*?" Certainly not. If Darwin's theory is true, we should leave it alone.

However, good science is now showing the theory wrong. That's a strong statement. A Darwinist might say something like: "Put your money where your mouth is!" Let's do it.

CHAPTER 2

Evidence Against the Theory of Evolution

Evidence from Genetics

Darwin's theory says fish evolved, through many intermediate steps, into human beings. The question thus arises: How did fish acquire the genes to become humans? A creature cannot be anything physically its genes won't allow. A zebra cannot give birth to a baby kangaroo—it only has zebra genes. A woman can't even be born blonde without genes for blonde hair—otherwise, she has to use Miss Clairol.

Genetics was not developed as a science in Darwin's day, and he assumed animals essentially had an unlimited capacity to adapt to environments.

He wrote: "By this process long continued . . . it seems to me almost certain that an ordinary hoofed quadruped might be converted into a giraffe."[13] In other words, Darwin believed you could take, say, donkeys, and if you put them in the right environment, they could, given enough time, become giraffes. This simply is not true. Even after millions of years in the jungle, donkeys would still be donkeys, because they only have donkey genes.

To resolve this dilemma, modern evolutionists asserted that the fish's genes must have *mutated* into human genes over eons—mutations, of course, are abrupt alterations in genes. They generally occur very rarely. According to evolutionary theory, an organism develops some new positive characteristic through a mutation, better adapting it to the environment. The creature then passes this mutated trait on to the next generation, and eventually it spreads through the whole species. Organisms without the trait, being weaker, die out ("survival of the fittest"). Through this process, fish gradually evolved into men.

However, this hypothesis no longer holds up. Dr. Lee Spetner, who taught information theory for a decade at Johns Hopkins University and the Weizman Institute, spent years studying mutations. He has written an important new book, *Not by Chance: Shattering the Modern Theory of Evolution.* In it, he writes: "In all the reading I've done in the life-sciences literature, I've never found a mutation that added information. . . . All point mutations that have been studied on the molecular level turn out to reduce the genetic information and not increase it."[14]

Mutations *delete* information from the genetic code. They never create higher, more complex information. What are they actually observed to cause in human beings? Death. Sterility. Hemophilia. Sickle cell anemia. Cystic fibrosis. Down's syndrome. And over 4,000 other diseases. The genetic code is designed to run an organism perfectly—mutations delete information from the code, causing birth defects.

To advance their view, evolutionists have long pointed to mutations with beneficial effects. The most common example given: mutations sometimes make bacteria resistant to antibiotics (germ-killing drugs). And so, the argument goes, "If mutations can make bacteria stronger, they must be able to do the same for other creatures." Dr. Spetner points out that this is based on a misunderstanding, for the mutations that cause antibiotic resistance still involve *information loss*.

For example, to destroy a bacterium, the antibiotic streptomycin attaches to a part of the bacterial cell called ribosomes. Mutations sometimes cause a structural deformity in ribosomes. Since the antibiotic cannot connect with the misshapen ribosome, the bacterium is resistant. But even though this mutation turns out to be beneficial, it still constitutes a *loss* of genetic information, not a gain. No "evolution" has taken place; the bacteria are not "stronger." In fact, under normal conditions, with no antibiotic present, they are weaker than their nonmutated cousins.

Let's take an analogy. Suppose a country's dictator ordered dissidents to be rounded up and handcuffed. So the police were busy handcuffing dissidents. But one day, they ran into a man born deformed—with no arms. One could conceivably say that, in this case, the man had an advantage over others, since he couldn't be handcuffed. But it certainly wouldn't represent an evolutionary advance. And neither does a deformity that prevents bacteria from being "handcuffed" by an antibiotic.

It is often possible to deduce a benefit from information loss. Suppose you ripped the windshield wipers off your car. Any benefit? Yes, your windshield could never be scratched by the wipers. But don't we all prefer wipers? Or suppose we just did away with cars completely. That would be a huge loss of information and technology, but there would be benefits: less pollution, and no one would die in car accidents.

What if a mutation causes a child to be born deaf? Any benefit? Yes, the child will never hear any curse words. But don't we all want children who can hear? In the same way, evolutionists, by viewing a particular mutation *in a limited context*, may describe the mutation as "beneficial" and incorrectly say it represents evolutionary progress. A good example is the disease sickle cell anemia, which some evolutionists have portrayed as beneficial because its deformed red blood cells are immune to malaria. But this is akin to saying it would be good to cut off your toes to prevent athlete's foot. Like the armless man, the wiperless car and the deaf child, these "beneficial mutations" turn out to be information losses.

Why is this a problem for evolution? Because if Darwin's thesis is correct, and all life began as a single cell, then chance mutations must have designed and engineered nearly every biological feature on Earth, from dolphins' remarkable sonar system (which is the envy of the U.S. Navy) to the human heart. The latter is an ingenious structure. Blood is pumped from the right side of the heart to the lungs, where it receives oxygen; back to the heart's left side, which propels it to the rest of the body through more than 60,000 miles of vessels. The heart has four chambers; a system of valves prevents backflow into any of these; electrical impulses from a natural pacemaker control the heart's rhythm.

Rarely, mutations cause babies to be born with congenital heart disorders, making blood shunt to the wrong place. There is no known case of mutations improving circulation. Hemoglobin—the blood's oxygen-carrying component—has over forty mutant variants. Not one transports oxygen better than normal hemoglobin.[15] To accept evolution, we must believe that human blood circulation—a wonder of engineering—was constructed by chance mutations, when actual observation demonstrates they *damage* it.

Ernst Chain, who shared a Nobel Prize for his work in developing penicillin, knew much about bacteria and antibiotics. Dr. Chain stated: "To postulate that the development and survival of the fittest is *entirely* a consequence of chance mutations, or even that nature carries out experiments by trial and error through mutations in order to create living systems better fitted to survive, seems to me a hypothesis based on no evidence and irreconcilable with the facts."[16]

Mutations *are* often inheritable, they *do* create changes, but the changes are inevitably *downward*, or at best neutral. Mutations have never been observed to originate a new hormone, organ, or other functional structure. They reduce, but do not generate, biologic technology. This is not to say it is *impossible* that a random mutation could create higher genetic information—only that it is not observed in science. *And Darwin's theory could die on this alone.* But instead, we'll just call this "strike one" on Darwin.

Evidence from Origins Science

Even larger difficulties arise with the evolutionary idea of life's beginnings. Charles Darwin and his contemporaries thought cells were rather simple, and that it would thus be feasible for chemicals in a "primordial soup" to come together and form one. However, through advances in biology, we now know that even a "simple" cell contains enough information to fill a hundred million pages of the *Encyclopaedia Britannica.*[17]

Cells consist essentially of proteins; one cell has thousands of proteins, and proteins are in turn made of smaller building blocks called amino acids. Normally, chains of hundreds of amino acids compose a protein, and these amino acids must be in precise functional sequence.

According to the evolutionary scenario, then, how did the first cell happen? Supposedly, amino acids formed in the primordial soup. Almost every high school biology text recounts Dr. Stanley Miller's famous experiment. In 1953, Miller, then a University of Chicago graduate student, assembled an apparatus in which he combined water with hydrogen, methane, and ammonia (proposed gases of the early Earth). He subjected the mixture to electric sparks. After a week, he discovered that some amino acids had formed in a trap in the system. Even though an ancient ocean would have lacked such an apparatus, evolutionists conjecture that in the primitive Earth, lightning (corresponding to Miller's electricity) could have struck a similar array of chemicals and produced amino acids. Since millions of years were involved, eventually they came, by chance, into the correct sequences. The first proteins were formed and hence the first cell.

But Sir Francis Crick, who shared a Nobel Prize for co-discovering DNA's structure, has pointed out how impossible that would be. He calculated that the probability of getting just one protein by chance would be one in ten to the power of 260—that's a one with 260 zeroes after it.[18] To put this in perspective, mathematicians usually consider anything with odds worse than one in 10 to the power of 50 to be, for practical purposes, impossible. Thus chance couldn't produce even one protein—let alone the thousands most cells require.

And cells need more than proteins—they require the genetic code. A bacterium's genetic code is far more complex than the code for Windows 98. Nobody thinks the program for Windows 98 could have arisen by chance (unless their hard drive blew recently).

But wait. Cells need more than the genetic code. Like any language, it must be translated to be understood. Cells have devices which actually translate the code. To believe in evolution, we must believe that, by pure chance, the genetic code was created, and also by pure chance, translation devices arose which took this meaningless code and transformed it into something with meaning.

Evolutionists cannot argue that "natural selection would have improved the odds." Natural selection operates in living things—here we are discussing dead chemicals that *preceded* life's beginning.

How could anything as complex as a cell arise by chance? A famous evolutionary argument dates to 1860, the year after publication of *The Origin of Species*. At Oxford University, "Darwin's bulldog" Thomas Huxley (whom we quoted earlier) engaged in a creation–evolution debate with theologian Samuel Wilberforce. There is no transcript, but reportedly Huxley, in making his case for chance origins, said that six monkeys, poking randomly at typewriters, and given enough millions of years, could write all the books in the British Museum. More than a century later, as a public school student, I heard a variation on that theme: "If a roomful of monkeys were to randomly clack at typewriters long enough, they would eventually recreate the complete works of Shakespeare. And if monkeys could recreate the complete works of Shakespeare by chance, then obviously a cell's information content could also arise by chance, if only given enough time."

However, anyone who believes these projections hasn't figured the math. What are the odds of a monkey typing one predetermined nine-letter word, such as "evolution"? We'll give Huxley a break, and assume a typewriter with only letters, no other symbols. Obviously, the first letter, "e," would be a piece of cake. But to get "evolution," since the alphabet has 26 letters, one must multiply 26 by itself eight times. We find the monkey would need, on average, more than five trillion attempts just to write "evolution" once correctly. Typing ten letters per minute, this would take over a million years. To get two *consecutive* predetermined nine-letter words—such as "evolution commenced"—would take more than a billion billion years, taking us much further back than the Big Bang, which supposedly occurred some 15 billion years ago. In other words, if a monkey started typing at the time of the Big Bang and continued until now, he couldn't even produce two consecutive preselected 9-letter words—let alone "the works of Shakespeare."

If it is objected that the example had a *roomful* of monkeys, Dr. Duane Gish puts the monkey matter in perspective:

> If one billion planets the size of the earth were covered eyeball-to-eyeball and elbow-to-elbow with monkeys, and each monkey was

> seated at a typewriter (requiring about 10 square feet for each monkey, of the approximately 10^{16} square feet available on each of the 10^{9} planets), and each monkey typed a string of 100 letters every second for five billion years, the chances are overwhelming that not one of these monkeys would have typed the sentence correctly! Only 10^{41} tries could be made by all these monkeys in that five billion years. . . . There would not be the slightest chance that a single one of the 10^{24} monkeys (a trillion trillion monkeys) would have typed a preselected sentence of 100 letters (such as "The subject of this *Impact* article is the naturalistic design of life on the earth under assumed primordial conditions") without a spelling error, even once.[19]

Even if the correct chemicals did come together by chance, would that create a living cell? Throwing sugar, flour, oil and eggs on the floor doesn't give you a cake. Tossing together steel, rubber, glass and plastic doesn't give you a car. These end products require skillful engineering. How much more so, then, a living organism? Indeed, suppose we put a frog in a blender and turn it into puree?

All the ingredients for life would be there—but nothing living arises from it. Even scientists in a lab can't produce a living creature from chemicals. How, then, could blind chance?

But let's say that somehow, by chance, a cell really formed in a primeval ocean, complete with all the necessary proteins, amino acids, genetic code, translation devices, a cell membrane, etc. Presumably this first little cell would have been rather fragile and short-lived. But it must have been quite a cell—because within the span of its lifetime, it must have evolved the complete process of cellular reproduction. Otherwise, there never would have been another cell.

And where did *sexual* reproduction come from? Male and female reproductive systems are quite different. Why would nature evolve a male reproductive system? Until it was fully functional, it would serve no purpose—and would *still* serve no purpose unless there was, conveniently available, a *female* reproductive system—which must also have arisen by chance.

Furthermore, suppose there really were some basic organic compounds formed from the "primordial soup." If free oxygen was in the atmosphere, it would oxidize many of those compounds —in other words, destroy them. To resolve this dilemma, evolutionists have long hypothesized that the Earth's ancient atmosphere had no free oxygen. For this reason, Stanley Miller did not include oxygen among the gases in his experiment.

However, geologists have now examined what they believe to be the Earth's oldest rocks and—while finding no evidence for an amino acid-filled "primordial soup"—have concluded that the early Earth was probably rich in oxygen.[20] But let's say the evolutionists are right—the early Earth had no free oxygen. Without oxygen, there would be no ozone, and without the ozone layer, we would receive a lethal dose of the sun's radiation in just 0.3 seconds.[21] How could the fragile beginnings of life have survived in such an environment?

Although we have touched on just a few of the problems of "chemical evolution," we can see that the hypothesis is, at every step, effectively impossible. Yet today, even first-grade children are taught the "fact" that life began in the ancient ocean as a single cell—with the scientific obstacles almost never discussed.

Darwin's theory could also die on this information alone, but instead we'll just call it "strike two."

Evidence from Biochemistry

Biochemistry is also giving Darwin problems. Dr. Michael Behe, biochemist at Lehigh University, has written *Darwin's Black Box: The Biochemical Challenge to Evolution*. In this 1996 book, Behe describes how complex certain biochemical systems are. If any component was missing, the system would have no function. Therefore it could not have evolved step-by-step. Behe calls this "irreducible complexity."

For example, blood clotting swings into action when we get a cut. A clot may look simple to the naked eye. However, through a microscope, it is a very complex process involving more than a dozen steps. A person with hemophilia is missing just one clotting factor and is at high risk for bleeding. Someone missing several components would have no chance for survival at all. To paraphrase Dr.

Behe very simply, if blood clotting had evolved step-by-step over eons, creatures would have bled to death before it was ever perfected—and its incremental stages never passed on to subsequent generations. The system is irreducibly complex.[22]

Another example Behe gives: the immune system. When infections occur, it must distinguish invading bacterial cells from the body's own cells—otherwise the latter will be attacked (which is the case in "autoimmune" diseases). An antibody identifies the bacterium by attaching to it. In a complex biochemical process, a variety of white blood cells—"killer cells"—are notified of the bacterium's presence. These travel to the site, and, using the identifying antibody, attack the enemy.

Like blood clotting, this system is irreducibly complex. The parts are interdependent. What evolved first? The killer cells? Without the identifying antibody, they wouldn't know where to attack. But why would the identifier develop first, without killer cells to notify? And if the network evolved gradually, disease would have wiped creatures out long before it could have been perfected.

Behe demonstrates that other biochemical systems, such as human vision, are also irreducibly complex—they cannot have evolved step-by-step—giving clear evidence that they resulted from intelligent design.

By my count, this puts three strikes on Darwin. But let's say this last strike was only a foul ball strike—we'll keep him at the plate.

Evidence from Fossils

Does paleontology—the study of fossils—validate evolution? Open a teenager's biology textbook, and you will probably see a "tree of life" from which all life forms branch out. At the tree's bottom is a single-celled creature. According to Darwinism, this little organism gradually evolved into the first invertebrates (creatures without backbones, such as jellyfish).

Cambrian rock is the low geologic layer containing most of the oldest known invertebrate fossils. In it, we find literally billions of fossils of invertebrates: clams, snails, worms, sponges, jellyfish, sea urchins, swimming crustaceans, etc. But there are no fossils demonstrating how these creatures evolved, or that they developed from a common ancestor. (For this reason, we hear of the Cambrian "explosion.") The late Stephen Jay Gould of Harvard acknowledged that "our more extensive labor has still failed to identify any creature that might serve as a plausible immediate ancestor for the Cambrian faunas [animals]."[23] In other words, the bottom of Darwin's great "tree of life" is merely a speculation unsupported by fossil evidence.

Supposedly, invertebrates evolved into the first fish. But despite billions of fossils from both groups, transitional fossils linking them are missing.

All through the evolutionary tree, the "missing links" are still missing. Insects, rodents, pterodactyls, palm trees and other life forms appear in the fossil record with no trace of how they evolved. Gareth J. Nelson of the American Museum of Natural History stated: "It is a mistake to believe that even one fossil species or fossil 'group' can be demonstrated to have been ancestral to another."[24]

Colin Patterson, senior paleontologist at the British Museum of Natural History, wrote: "Gould and the American Museum people are hard to contradict when they say there are no transitional fossils. As a paleontologist myself, I am much occupied with the philosophical problems of identifying ancestral forms in the fossil record. . . . I will lay it on the line—there is not one such fossil for which one could make a watertight argument."[25]

Many other paleontologists have made equally strong affirmations (see Chapter 2 of my book *Tornado in a Junkyard*). Of course, this certainly does *not* mean that there are no transitional forms claimed today by evolutionists. Indeed, with the rising challenge to Darwin, fewer evolutionists seem to acknowledge the lack of transitional fossils, perhaps for fear of being quoted by creationists. Some have begun to more strongly assert the existence of such forms.

However, the vulnerability of such opinions to error is demonstrated by times when they have been conclusively proven wrong. Take, for example, Piltdown Man. It was declared an ape-man,

500,000 years old. It was validated by many of Britain's leading scientists, including noted anatomist Sir Arthur Keith, brain specialist Sir Grafton Eliot Smith, and British Museum geologist Sir Arthur Smith Woodward. At the time the discovery was announced (1912), the *New York Times* ran this headline: "Darwin Theory Proved True."[26] For the next four decades, Piltdown Man was evolution's greatest showcase, featured in textbooks and encyclopedias. Meanwhile, clergymen who had denounced evolution were ridiculed; Piltdown, it was said, had proven them wrong.

But what did the Piltdown Man actually consist of? Just a very recent orangutan jaw, stained to look old, its teeth filed down to make them more human-looking, planted together with a human skullbone, also stained to create an appearance of age.

Those who think such mistakes no longer occur need only consider the *Archaeoraptor*, promoted in a 10-page color spread in the November 1999 *National Geographic* as a "true missing link" between dinosaurs and birds. The fossil was displayed at National Geographic's Explorers Hall and viewed by over 100,000 people. However, it too turned out to be a fake—someone had simply glued together a bird fossil with part of a dinosaur fossil.

Nor is it just fraud that can deceive. The coelacanth is a bony fish whose fossils can be seen in Jurassic rock (the age of the dinosaurs). Supposedly this creature had been extinct for some 70 million years. According to Darwin's theory, fish evolved into amphibians (animals that can go on land and water, such as frogs). For years, evolutionists called the coelacanth a forerunner of amphibians, its fossilized fins described as limb-like.

Then, in 1938, fishermen caught a live one off the African coast. Since then, about 200 more have been caught. Besides proving the coelacanth was not extinct for 70 million years, examination revealed it was 100 percent fish, with no amphibian characteristics.

Why is it relatively easy to be misled by a fossil? Since 99 percent of an organism's biology resides in its soft anatomy, there is a limit to how much one can deduce from a bone. This makes fossils easy to invest with subjective opinions. As Jerold Lowenstein and Adrienne Zihlman noted in *New Scientist*, in reference to human ancestry:

> The subjective element in this approach to building evolutionary trees, which many palaeontologists advocate with almost religious fervor, is demonstrated by the outcome: there is no single family tree on which they agree.[27]

There is no conclusive way to test the interpretation of the fossil of an extinct creature. Science cannot observe the past with the same authority as it observes the present. Paleontology, therefore, is not a science on the level of physics or chemistry, whose laws can be demonstrated in a laboratory. It relies heavily on *opinion* and might even better be described as an art than a science.

On the 25th anniversary of President John F. Kennedy's death, a national magazine asked me to write an in-depth article on the assassination. In researching it, I was astonished at the variety of opinions about what had occurred—the identity of the assassin(s), number of assassins, locations from which they had fired, etc. These debates raged despite a wealth of evidence: hundreds of eyewitnesses interviewed by the Warren Commission; the Zapruder film which caught the actual slaying; fingerprints; ballistics tests. Even the autopsy results on Kennedy's body were disputed in a best-selling book.

If this much debate can occur over an incident that happened only 40 years ago, how then can an evolutionist pick up a bone fragment, supposedly millions of years old, and assert with *a high degree of certainty* that it is the ancestor of such-and-such a species? Unlike the Kennedy assassination, there are no eyewitnesses who saw this creature; there is no Zapruder movie of it; there are no soft tissues to examine.

Darwin stated that "the number of intermediate and transitional links, between all living and extinct species, must have been inconceivably great. But assuredly, if this theory be true, such have lived upon the earth."[28] He admitted these creatures' fossils had not been found in his day, but hoped future excavations would turn them up. They haven't.

If evolutionary theory is true, the geologic record should reveal the innumerable transitional forms Darwin spoke of. We shouldn't find just a handful of questionable fossils, but billions of intermediates validating his theory. Instead, the fossil record shows animals complete—not in developmental stages—the very first time they are seen. This is just what we would expect if animals were created, instead of evolved.

This is another strike on Darwin, but since the subjectivity of the fossil record makes it a more debatable issue, we'll call it another foul ball strike. Darwin can stay at the plate.

Evidence from Taxonomy

What about *living* transitional forms? Taxonomy is the science that classifies plants and animals, grouping them by characteristics they share. Swedish botanist Carolus Linnaeus pioneered the field, assigning organisms by class, order, genus and species. His system won universal acceptance. Linnaeus strongly opposed evolution. He saw that the larger divisions of living things—contrary to what evolution would predict—were distinctly divided without overlaps.

A rainbow may have many colors, but one doesn't see solid red jump to solid orange. Rather, gradations exist between them. Similarly, if all creatures have a common ancestor, we should not see distinctly divided groups, but living intermediates between those groups. Evolutionists acknowledge that the intermediates are missing, but say they must have become extinct. But if so, where are their fossils? Canadian biologist W. R. Thompson noted:

> Taking the taxonomic system as a whole, it appears as an orderly arrangement of clear-cut entities, which are clear-cut because they are separated by gaps. . . . The general tendency to eliminate, by means of unverifiable speculations, the limits of the categories nature presents to us, is the inheritance of biology from the *Origin of Species*. To establish the continuity required by the theory, historical arguments are invoked, even though historical evidence is lacking. Thus are engendered those fragile towers of hypotheses based on hypotheses, where fact and fiction intermingle in an inextricable confusion.[29]

Little has changed since 1930, when Austin H. Clark, the Smithsonian Institution's eminent zoologist, declared:

> The complete absence of any intermediate forms between the major groups of animals, which is one of the most striking and most significant phenomena brought out by the study of zoology, has hitherto been overlooked, or at least ignored. . . .
>
> No matter how far back we go in the fossil record of previous animal life upon the earth we find no trace of any animals forms which are intermediate between the various major groups or phyla.
>
> This can only mean one thing. There can be only one interpretation of this entire lack of any intermediates between the major groups of animals—as for instance between the backboned animals or vertebrates, the echinoderms, the mollusks and the arthropods.
>
> If we are willing to accept the facts we must believe that there never were such intermediates, or in other words that these major groups have from the very first borne the same relation to each other that they bear today.[30]

Evidence from Molecular Biology

According to Darwinism, fish evolved into amphibians, which then evolved into reptiles, which then evolved into mammals. Australian molecular biologist Michael Denton studied these different animals on a molecular level, and found no evidence for the sequence. In his book, *Evolution: A Theory in Crisis*, Denton analyzes various molecular structures, such as that of cytochrome C, a protein involved in producing cellular energy. It is found in organisms ranging from bacteria to man. Based on cytochrome C, amphibians are just as distant from fish as people are. In other words, on a molecular level, amphibians are *not* close cousins of fish. Denton writes:

> Instead of revealing a multitude of transitional forms through which the evolution of a cell might have occurred, molecular biology has served only to emphasize the enormity of the gap. . . . [N]o living system can be thought of as being primitive or ancestral with respect to any other system, nor is there the slightest empirical hint of an evolutionary sequence among all the incredibly diverse cells on earth.

> . . . [T]he system of nature conforms fundamentally to a highly ordered hierarchic scheme from which all direct evidence for evolution is emphatically absent.[31]

Issues of Common Sense

In a popular evolutionary explanation, here's how reptiles evolved into birds: They wanted to eat flying insects that were out of reach. So the reptiles began leaping, and flapping their arms to get higher. Over millions of years, their limbs transformed into wings by increments, their tough reptilian scales gradually sprouting soft feathers.

But the theory suffers when scrutinized. A few years ago, I was walking through a zoo with my son. We saw uncaged parrots sitting on perches out in the open. My son asked me why the parrots didn't just fly away. We queried the zookeeper, who told us: "We clip their wings."

Now, what would happen to these parrots if turned loose in the jungle? Unable to fly, they would make easy targets for predators and swiftly perish.

According to natural selection, a physical trait is acquired because it enhances survival. Obviously, flight is beneficial. One can certainly see how flying animals might survive better than those who couldn't, and thus natural selection would preserve them. But birds' wings and feathers are perfectly designed instruments. "Evolving" wings would have no genuine survival value until they reached the point of flight. The transitional creature whose limb was half leg, half wing, would be a poor candidate for survival—it couldn't fly yet, nor walk well. Natural selection would eliminate it without a second thought.

Let's raise an even more fundamental question: Why aren't reptiles *today* developing feathers? Why aren't fish *today* growing little legs, trying to adapt to land? Why aren't invertebrates evolving into vertebrates? Why aren't reptiles evolving into mammals? Shouldn't evolution be ongoing?

And why is man so incredibly different from animals? What animal can solve complex math equations? Write poetry? Laugh at jokes? Design computer software? How can we say man is merely "one more animal, just more highly evolved"?

Let's review the evidence against Darwin's theory:

(1) Mutations, the supposed building blocks of evolution, are never actually observed to create higher genetic information.

(2) Cells are far too complex to have originated from a chance arrangement of chemicals.

(3) The human body has features, such as blood clotting and the immune system, that are "irreducibly complex" and cannot have evolved.

(4) The fossil record reveals animals complete when first seen and thus better supports creation than evolution.

(5) Taxonomy shows a lack of intermediates between the major divisions of living creatures.

(6) On a molecular level, there is no evidence for the evolutionary sequence.

(7) Common sense argues against evolution.

This is probably enough to conclusively call "strike three" on the theory. But first let's examine various evidences used to *support* Darwinism.

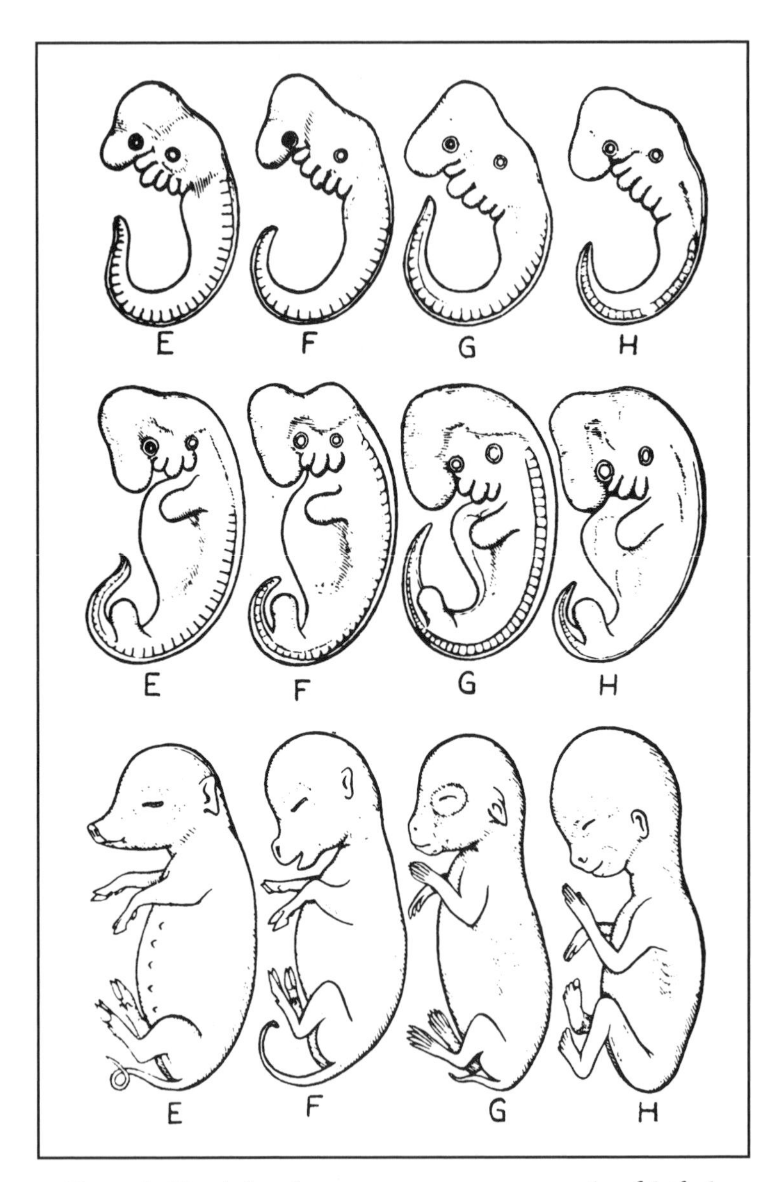

Figure 1. Haeckel embryo sequence, now exposed as falsified, purported to show (left to right) hog, calf, rabbit, human.

CHAPTER 3

Reevaluating Some Evidences Used to Support the Theory

Haeckel's Embryos

Most of us have seen those drawings of developing human embryos next to developing animal embryos, and they look virtually indistinguishable (Figure 1). This has long been said to demonstrate our common ancestry with these animals and thus prove the theory of evolution.

These pictures were designed by German zoologist Ernst Haeckel (whom we previously quoted on the "lower races" having a "different value to their lives"). Haeckel explained:

> When we see that, at a certain stage, the embryos of man and the ape, the dog and the

> rabbit, the pig and the sheep, though recognizable as higher vertebrates, cannot be distinguished from each other, the fact can only be elucidated by assuming a common parentage.[32]

What few people know: the pictures were fakes. The deceit was exposed in *Haeckel's Frauds and Forgeries* (1915), a book by J. Assmuth and Ernest R. Hull. They quoted nineteen leading authorities of the day. Anatomist F. Keibel of Freiburg University said that "it clearly appears that Haeckel has in many cases freely invented embryos, or reproduced the illustrations given by others in a substantially changed form."[33] Zoologist L. Rütimeyer of Basle University called the distorted drawings "a sin against scientific truthfulness."[34]

Despite the early exposure, Western educators continued using Haeckel's pictures for decades as proof of the theory of evolution. The matter has been settled with finality by Dr. Michael Richardson, an embryologist at St. George's Medical School, London. He found there was no record that anyone ever actually checked Haeckel's claims by systematically comparing human and other fetuses during development. He assembled a scientific team that did just that—photographing the growing embryos of 39 different species. In a 1997 interview in London's *The Times*, Dr. Richardson stated: "This is one of the worst cases of scientific fraud. It's shocking to find that somebody one thought was a great scientist was deliberately misleading. It makes me angry. . . . What he [Haeckel] did was to take a human embryo and copy it, pretending that the salamander and the pig and all the others looked the same at the same stage of development. They don't. . . . These are fakes."[35]

Today, though to a lesser extent, Haeckel's drawings still appear in a number of high school and college textbooks.

Vestigial Organs

In 1925, evolutionary zoologist Horatio Hackett Newman stated: "There are, according to Wiedersheim, no less than 180 vestigial structures in the human body, sufficient to make of a man a veritable walking museum of antiquities."[36] This was another of Darwinism's great myths: the human body is loaded with vestigial organs—relics from our animal past no longer serving any significant purpose.

One reason why so many tonsillectomies were previously performed was the false belief that tonsils were "vestigial." Today the tonsils are recognized as having an immune function. Evolutionists said the pineal gland, located in the brain, was vestigial—now we know it secretes the hormone melatonin. The thymus, found in the chest, was also declared useless. We have since discovered it has an immune function. The thyroid, coccyx, and many other body parts previously deemed "vestigial" are now understood to have important uses. The list of 180 vestigial structures is practically down to zero. Unfortunately, earlier Darwinists assumed that ignorance of an organ's function meant it *had* no function.

Salt

In 1962, President John F. Kennedy, speaking at a dinner for the America's Cup crews, stated that "we all came from the sea." He then repeated a popular misconception: "And it is an interesting biological fact that all of us have, in our veins, the exact same percentage of salt in our blood that exists in the ocean." Kennedy went on to say people enjoy sailing because "we are going back to from whence we came."[37]

This erroneous idea—that our blood contains "the exact same percentage" of salt as the ocean—was widespread. It was stated by my sixth-grade teacher, and seemed impressive evidence that man had evolved from sea creatures.

However, human blood does not resemble sea water. The actual mineral content of human blood plasma and seawater, in milligrams per liter, is as follows:

ELEMENT	BLOOD	SEAWATER
Sodium	3200	10800
Chlorine	3650	19400
Potassium	200	392
Calcium	50	411
Magnesium	27	1290
Phosphorus	36	0.09
Iron	1	0.004
Copper	1	0.001
Zinc	1.1	0.005
Chromium	1.1	0.0002

Bromine	4	67
Fluorine	0.1	1.3
Boron	1	5
Selenium	0.9	0.0001[38]

Thus another claim—like Haeckel's embryos and Huxley's typing monkeys—was perpetuated because no one bothered to check the facts.

Human Tails

Another long-held idea was that some babies are born with "tails"—throwbacks to the days when we were apes swinging from trees. A. Rendle Short, professor of surgery at the University of Bristol, clarified this long ago:

> It is often stated that children are born with "tails"; but as a rule the alleged "tails" are nothing but fatty or fibrous tumors such as may be met with in many parts of the body, without any embryological significance. . . . There are many congenital abnormalities with which the medical profession is well acquainted: club foot, hare lip, cleft palate, congenital dislocations, naevi, supernumerary fingers and toes, *spina bifida*. But none of these recall the ape.[39]

Peppered Moths

The latest Darwinian evidence to bite the dust: the peppered moth. Most high school biology textbooks today cite this creature as a proof of evolution. Britain's peppered moth comes in light and dark varieties. According to the evolutionary scenario, the peppered moth rests on tree trunks during the day. Industrial pollution blackened the trunks, making the dark moths invisible to preying birds, which caused them to become the dominant variety. Later, pollution controls caused the light variety to resurge.

Biologist Jonathan Wells, who holds PhDs from Yale and Berkeley, exposes the myth of the peppered moth in his 2000 book *Icons of Evolution*. As it turns out, peppered moths do *not* rest on tree trunks during the day. In some studies, this was faked by gluing and pinning dead moths to trees and photographing them.[40] Even if the studies had been completely valid, all they would have shown is fluctuation within a species. There is no question that such fluctuations occur. The real question is whether one kind of animal (fish) can become a completely different kind (mammal)—that's what Darwin's theory is all about.

With so many Darwinist proofs crumbling, what remains? Two major arguments undergird evolution today.

The Argument from "Microevolution"

Modern evolutionists, like Darwin himself, use breeding experiments as evidence. We are reminded that dog breeders have developed new breeds of dog; that racehorse owners have bred faster horses; that horticulturists have developed new plant varieties; etc. This is said to show that living things change over time. Therefore, given *lots* of time, *lots* of change would occur, and over *unlimited* time, *unlimited* change would occur—i.e., fish to human—since nature conducts its own form of "breeding" by allowing only the fittest creatures to survive.

The argument is flawed, however. Let's take dog breeding. Were the dog breeders of past centuries genetic scientists? Did they sit in labs inserting new genes into these dogs? No. They found dogs that already had characteristics they liked, mated them with similar dogs, and bred them in a certain direction. In other words, they worked with *preexisting* genetic information.

A species is normally endowed with a rich, diverse gene pool. Take man himself. There are over six billion humans on Earth, yet no two are exactly alike. (If one wanted to get technical, it could be argued that identical twins are alike.) The human race has a vast gene pool that permits all the variation we see.

It certainly is possible to change the general appearance of a species over time, by selecting out creatures with particular genes. But the change is confined to the limits of the gene pool. Horse breeders can generate fast horses by choosing the best, but they cannot convert the horse into a different animal.

In nature's parallel, if a group of frogs flee a forest fire, perhaps only the fastest hoppers would escape. This could leave us with a strain of fast frogs. It's a perfect example of "natural selection" and "survival of the fittest." But this doesn't mean the frogs could evolve into people. Nothing new has been created. All that has happened is that *slower frogs have been eliminated*—this is, in fact, a loss of genetic information, not a gain. We would not contest Darwin on the existence of natural selection or survival of the fittest in nature. But we would argue that the change possible *does* have limits. (One comic pointed out: "The princess kissed the frog, and he turned into a handsome prince. We call that a fairy tale. Darwin says frogs turn into princes, and we call it science.")

Shifts can and do occur *within* types of animals. These changes, based on diverse, preexisting genetic information, are called by some "microevolution." But this is not evidence for *unlimited* transformation ("macroevolution").

The thesis "if we get a little change over a little time, then we will get a lot of change over a lot of time" does not hold up. Suppose a girl, dreaming of Olympic glory, learns to ice skate. The first week, she finds she can jump to a height of one foot and land on her skates. The second week, she finds that she can leap two feet. The third week, she can jump three feet. Can we conclude from this that after 100 weeks, she will be able to jump

100 feet? No, the law of gravity will strictly limit how high she can get. Likewise, animals are also restricted in how much change *they* can make—by the limits of their gene pool.

Luther Burbank, the famed American plant breeder, said:

> I know from my experience that I can develop a plum half an inch long or one 2½ inches long, with every possible length in between, but I am willing to admit that it is hopeless to try to get a plum the size of a small pea, or one as big as a grapefruit. . . . I have roses that bloom pretty steadily for six months in the year, but I have none that will bloom twelve, and I will not have. In short, there are limits to the development possible, and these limits follow a law. . . . plants and animals all tend to revert, in successive generations, toward a given mean or average.[41]

Simple bacteria can produce another generation in a matter of minutes. Yet Alan H. Linton, emeritus professor of bacteriology at the University of Bristol, noted in 2001: "Throughout 150 years of the science of bacteriology, there is no evidence that one species of bacteria has changed into another."[42]

To change a bacterium into a fish into a frog into a reptile into a mammal into a man, would require that each type of creature largely rewrite its gene pool and replace it with a new one. Evolutionists contend that "beneficial mutations" would allow an animal to exceed the boundaries of its genetic makeup. But as we have seen, mutations do not introduce new genetic information; the changes they cause involve informational losses.

The Second Major Argument—Similarity

Suppose we're munching on a Big Mac at McDonald's and see a large family sitting next to us. We notice that the brothers and sisters resemble each other. Why is that? Because they have the same parents from whom they get their genes (traits). Charles Darwin also noticed this. He correctly deduced that traits are inherited from parents. However, once again, he stretched the conclusion.

The Darwinist looks at a human and a tiger. He notes that both creatures have two eyes, two ears, four limbs, a heart, brain, teeth, and so forth. The conclusion drawn: since people and tigers share so many characteristics, this proves a common biological ancestor. In other words, the man and the tiger are "brothers" having a common parent many generations ago. A child's biology textbook today will often display similarities between people and animals, as in their limb bones, as proof of a mutual line of descent.

However, similarities can be explained in other ways. Take a Toyota Camry and Ford Explorer. Both have headlights, a transmission, a windshield, four wheels, and hundreds of other features in common. Yet this did not result from chance mutations or any other biological process. Every component in an automobile is the fruit of deliberate planning.

Yes, similarities can derive from biologic ancestry. But they also result from the necessities of intelligent design. Cars have four wheels because that's the best arrangement (just try driving a car with less). In the same way, God may have created animals with four limbs because it was the *best design.*

A few evolutionists try to usurp the car analogy, saying all autos evolved from a basic prototype, and were modified over time, just as natural selection modified creatures over time. However, no one questions that *ideas* evolve. Ideas rapidly change, limited only by the bounds of our imagination. Changing a species, which is strictly confined to its gene pool, is another story. That ideas evolve does not argue for *biological* evolution.

Similarities can also result from the style of a *common designer*. A person with an eye for architecture can identify a house built by the famed Frank Lloyd Wright. But the similarity between his buildings doesn't mean one gave birth to another. They had a *common designer*. Man and animals may also have one: God.

We should also consider the claim that *genetic* similarities prove biological relationships. High school students today are often told that "genetically speaking, men and chimpanzees are 98 percent similar."

Most people, hearing this argument, assume scientists compared the genetic code sequences of humans and chimpanzees, and found them nearly identical. However, the genetic map of human beings was only recently laid out in the famous Celera project. The "98 percent" argument existed long before that; no one had systematically compared the genetic makeup of humans and chimpanzees. What, then, was the argument based on? It came from a process called "DNA hybridization," which found that a single strand of human DNA and a single strand of chimp DNA could rather conformably be put together to form a double strand.

However, suppose someone eventually does compare the genetic code sequences of chimps and humans and conclusively shows they are similar? Would this prove common ancestry? To answer that, let's take another analogy—the Internet.

There are millions of web pages, and it is probably correct to say that most were created using codes of the computer language HTML. Does having these codes in common imply a genealogic relationship between web pages? Did the web site for radio station WEZE in Boston give birth to the web site for WAVA in Virginia? No. Every web site results from intelligent design, and the computer codes *used to make them* resulted from intelligent design.

Two similar-looking web pages probably have similar HTML code commands. By the same token, two similar-looking animals would likely have some corresponding genetic code commands. Since humans resemble chimps more than whales, we could expect that human genetic code sequences would be closer to that of chimps than that of whales. But no biological relationship would be thereby certified.

Evolutionists argue that similarities, visible or genetic, *prove* a common ancestor. But this is an assumption. Similarities *can* result from common ancestry, but also from intelligent design or a common designer.

Conclusion

What, then, does evolution's proof consist of? Where ideas have not been discredited (as in Haeckel's embryos and vestigial organs), we have seen that they rest on *assumptions* rather than *observations*. No one has ever observed life spontaneously generate from chemicals, or one kind of animal transform into another, or mutations generate true biological advances, or complex biochemical systems evolve. That any of these things ever happened requires *faith* by the Darwinist, and for that reason, some people consider evolution better characterized as a religion than as a science.

Sir John William Dawson, who pioneered Canadian geology and served as president of both McGill University and the British Association for the Advancement of Science, said:

> Let the reader take up either of Darwin's great books, or Spencer's "Biology," and merely ask himself as he reads each paragraph, "What is assumed here and what is proved?" and he

> will find the whole fabric melt away like a vision. . . . We thus see that evolution as an hypothesis has no basis in experience or in scientific fact, and that its imagined series of transmutations has breaks which cannot be filled.[43]

It is common to hear it asserted that "all scientists believe in evolution." But many scientists, from Darwin's day until now, have rejected it. Zoologist Albert Fleischmann of the University of Erlangen declared: "The Darwinian theory of descent has not a single fact to confirm it in the realm of nature. It is not the result of scientific research, but purely the product of imagination."[44]

Paul Lemoine, who was president of the Geological Society of France and director of the Natural History Museum, Paris, stated:

> The theory of evolution is impossible. At base, in spite of appearances, no one any longer believes in it. . . . Evolution is a kind of dogma which the priests no longer believe, but which they maintain for their people.[45]

Dr. Wolfgang Smith, who taught at MIT and UCLA, and has written on a wide spectrum of scientific topics, said in 1988:

> And the salient fact is this: *if by evolution we mean macroevolution* (as we henceforth shall), *then it can be said with the utmost rigor that the doctrine is totally bereft of scientific sanction.* Now, to be sure, given the multitude of extravagant claims about evolution promulgated by evolutionists with an air of scientific infallibility, this may indeed sound strange. And yet the fact remains that there exists to this day not a shred of *bona fide* scientific evidence in support of the thesis that macroevolutionary transformations have ever occurred.[46]

In Australia in 1999, a book was published entitled *In Six Days: Why 50 Scientists Choose to Believe in Creation*. It has 50 chapters, each written by a scientist holding a doctorate, none of whom accepts Darwin's theory. In the United States, the Creation Research Society has some 600 voting members, all holding advanced science degrees, and all of whom reject Darwinian evolution. Dr. Raymond Damadian, inventor of the MRI—one of the most advanced diagnostic tools in medicine—is an outspoken creationist. Many more examples can be given—and are in my book *Tornado in a Junkyard*.

But of course, many scientists *do* accept evolution, and it would be logical to ask why that is, if evidence really goes against it. I believe there are two answers:

(1) Many scientists accept evolution because *that's all they've ever been taught*. Having never been exposed to the case against Darwin, they have never had a chance to weigh it.

(2) Another reason was well summarized by Dr. Michael Walker, senior lecturer in anthropology at Sydney University:

> One is forced to conclude that many scientists and technologists pay lip-service to Darwinian Theory only because it supposedly excludes a Creator from yet another area of material phenomena, and not because it has been paradigmatic in establishing the canons of research in the life sciences and the earth sciences.[47]

Like everyone else, scientists are human. As humans we often dislike moral laws or the idea of God. It is probably reasonable to say that Darwinism largely persists, despite contradictory evidence, because it is, at its base, a denial of God's existence. The grandson of Thomas Huxley—"Darwin's bulldog"—was Aldous Huxley, an early advocate of the drug culture and sexual permissiveness. He put it bluntly:

> I had motives for not wanting the world to have meaning; consequently assumed it had none, and was able without any difficulty to find satisfying reasons for this assumption. . . . For myself as, no doubt, for most of my contemporaries, the philosophy of meaninglessness was essentially an instrument of liberation. The liberation we desired was simultaneously liberation from a certain political and economic system and liberation from a certain system of morality. We objected to the morality because it interfered with our sexual freedom; we objected to the political and economic system because it was unjust.[48]

By all means, we should permit evolutionary ideas to be aired in public schools. But they should be stated for what they are: hypotheses, not absolutes. Charles Darwin confided his own doubts in a letter to a colleague in 1858, the year before publication of *The Origin of Species*:

> Thank you heartily for what you say about my book; but you will be greatly disappointed; it will be grievously too hypothetical. It will very likely be of no other service than collating some facts; though I myself think I see my way approximately on the origin of the species. But, alas, how frequent, how almost universal it is in an author to persuade himself of the truth of his own dogmas.[49]

If Darwin himself called it "grievously too hypothetical," why are we teaching it as proven fact today? Schools should also present the growing scientific case *against* Darwinism, so students can weigh both sides and make up their own minds. Of course, some regard any challenge to Darwin as "sneaking religion in via the back door," and try to silence teachers who present these evidences as "violating separation of church and state." But it is quite a stretch to claim that discussing, say, the failure of mutations to add new genetic information, somehow constitutes a government endorsement of religion. Many, in fact, consider evolution a way of sneaking *atheism* in through the back door—and all the ideology that goes with it.

Wernher von Braun was director of NASA's space flight center; he oversaw the team of scientists that sent the first American into space, and masterminded the moon landing. Regarding science education, he stated: "To be forced to believe only one conclusion—that everything in the universe happened by chance—would violate the very objectivity of science itself."[50]

Indeed, suppose an arson detective, about to investigate a fire's origins, received these instructions: "In your investigation, you must only consider the possibility that the fire happened by chance; you must not explore evidence that suggests it was intentional." How sound would that investigation be?

A few years ago, the National Academy of Sciences issued an *Affirmation of Freedom of Inquiry and Expression* that stated: "That freedom of inquiry and dissemination of ideas require that those so engaged be free to search where their inquiry leads, free to travel and free to publish their findings without political censorship and without fear of retribution in consequence of unpopularity of their conclusions. Those who challenge existing theory must be protected from retaliatory reactions."[51]

Science is about the truth, and it neither fears nor suppresses the search for it.

Notes

1. Julian Huxley, in *Issues in Evolution*, vol. 3, ed. Sol Tax (Chicago: University of Chicago Press, 1960), 45.
2. Emelian Yaroslavsky, *Landmarks in the Life of Stalin* (Moscow: Foreign Languages Publishing House, 1940), 8-9.
3. Conway Zirkle, *Evolution, Marxian Biology, and the Social Scene* (Philadelphia: University of Philadelphia Press, 1959), 86.
4. Andrew Carnegie, *Autobiography of Andrew Carnegie*, ed. John C. Van Dyke (1920; reprint, Boston: Northeastern University Press, 1986), 327.
5. Charles Darwin, *The Descent of Man and Selection in Relation to Sex* (New York: D. Appleton, 1896), 156.
6. Thomas Huxley, *Lay Sermons, Addresses and Reviews* (New York: Appleton, 1870), 20.
7. Darwin, *Descent of Man*, 133-34.
8. Ernst Haeckel, *The Wonders of Life* (New York: Harper, 1904), 56-57.
9. A. E. Wilder-Smith, *The Natural Sciences Know Nothing of Evolution* (Costa Mesa, Calif.: T.W.F.T. Publishers, 1981), 162.
10. Arthur Keith, *Evolution and Ethics* (New York: Putnam, 1947), 230.
11. E. O. Wilson, "Toward a Humanistic Biology," *The Humanist* (September/October 1982): 40.
12. Will Durant, "We Are in the Last Stage of a Pagan Period," *Chicago Tribune* (April 1980), quoted in

Henry M. Morris and John D. Morris, *Society and Creation* (Green Forest, Ark.: Master Books, 1996), 80.

13. Charles Darwin, *The Origin of Species* (1872; reprint, New York: Random House, 1993), 278.
14. Lee Spetner, *Not By Chance!: Shattering the Modern Theory of Evolution* (Brooklyn, N.Y.: Judaica Press, 1997), 131, 138.
15. Gary E. Parker, "Creation, Mutation, and Variation," *Impact* 89 (November 1980): 2.
16. Ernst Chain, *Responsibility and the Scientist in Modern Western Society* (London: Council of Christians and Jews, 1970), 25.
17. Carl Sagan, "Life," *Encyclopaedia Britannica*, 15th ed., vol. 22, 987.
18. Francis Crick, *Life Itself: Its Origin and Nature* (New York: Simon and Schuster, 1981), 51-52.
19. Duane T. Gish, "The Origin of Life: Theories on the Origin of Biological Order," *Impact* 37 (July 1976): 3.
20. Harry Clemmey and Nick Badham, "Oxygen in the Precambrian Atmosphere: An Evaluation of the Geological Evidence," *Geology* 10 (March 1982): 141.
21. Carl Sagan, "Ultraviolet Selection Pressure on the Earliest Organisms," *Journal of Theoretical Biology* 39 (April 1973): 195, 197.
22. Michael Behe, *Darwin's Black Box: The Biochemical Challenge to Evolution* (New York: The Free Press, 1996), 77-97.
23. Stephen Jay Gould, "A Short Way to Big Ends," *Natural History* 95 (January 1986): 18.

24. Gareth V. Nelson, "Origin and Diversification of Teleostean Fishes," *Annals of the New York Academy of Sciences* 67 (1969): 22.
25. Colin Patterson, letter to Luther D. Sunderland, 10 April 1979, quoted in Luther D. Sunderland, *Darwin's Enigma: Fossils and Other Problems* (San Diego: Master Books, 1988), 89.
26. "Darwin Theory Proved True," *New York Times,* 22 December 1912, C1.
27. Jerold M. Lowenstein and Adrienne L. Zihlman, "The Invisible Ape," *New Scientist* 120 (3 December 1988): 58.
28. Darwin, *Origin*, 408.
29. W. R. Thompson, introduction to *The Origin of Species*, by Charles Darwin (reprint, New York: Dutton, Everyman's Library, 1956), quoted in Henry M. Morris and John D. Morris, *Science and Creation* (Green Forest, Ark.: Master Books, 1996), 29.
30. Austin H. Clark, *The New Evolution: Zoogenesis* (Baltimore: Williams and Wilkin, 1930), 168, 189.
31. Michael Denton, *Evolution: A Theory in Crisis* (Bethesda, Md.: Adler and Adler, 1986), 249, 250, 278.
32. Ernst Haeckel, *The Riddle of the Universe at the Close of the Nineteenth Century*, trans. Joseph McCabe (New York: Harper and Brothers, 1900), 65-66.
33. J. Assmuth and Ernest R. Hull, *Haeckel's Frauds and Forgeries* (Bombay: Examiner Press, 1915), 26.
34. Ibid., 24.
35. "An Embryonic Liar," *The Times* (London), 11 August 1997, 14.

36. *The World's Most Famous Court Trial: Tennessee Evolution Case* (Dayton, Tenn.: Bryan College, 1990), 268.
37. *Public Papers of the Presidents of the United States: John F. Kennedy, 1962*, vol. 1, (Washington, D.C.: United States Government Printing Office, 1963), 684.
38. Don Batten, "Red-blooded Evidence," *Creation Ex Nihilo* 19 (March-May 1997): 24.
39. A. Rendle Short, "Some Recent Literature Concerning the Origin of Man," *Journal of the Transactions of the Victoria Institute* 67 (1935): 256.
40. Jonathan Wells, *Icons of Evolution: Science or Myth?* (Washington, D. C.: Regnery, 2000), 149.
41. Norman Macbeth, *Darwin Retried* (Boston: Gambit, 1971), 36.
42. Alan H. Linton, "Scant Search for the Maker," *The Times Higher Education Supplement*, 20 April 2001, 29.
43. William Dawson, *The Story of Earth and Man* (New York: Harper and Brothers, 1887), 330, 339.
44. John Fred Meldau, ed., *Witnesses Against Evolution* (Denver: Christian Victory Publishing, 1968), 13.
45. Henry M. Morris, *Men of Science—Men of God* (El Cajon, Calif.: Master Books, 1988), 84.
46. Wolfgang Smith, *Teilhardism and the New Religion* (Rockford., Ill.: Tan Books, 1988), 5-6.
47. Michael Walker, "To Have Evolved or to Have Not? That is the Question," *Quadrant* 25 (October 1981): 45.
48. Aldous Huxley, *Ends and Means: An Inquiry into the Nature of Ideals and into the Methods Employed for Their Realization* (New York: Harper and Brothers, 1937), 312, 316.

49. Charles Darwin, *More Letters of Charles Darwin*, ed. Francis Darwin, vol. 1 (London: John Murray, 1903), 450.
50. Wernher Von Braun, letter read by Dr. John Ford to California State Board of Education, 14 September 1972, quoted in Ann Lamont, *Twenty-One Great Scientists Who Believed the Bible* (Acacia Ridge, Queensland, Australia: Creation Science Foundation, 1995), 47.
51. "An Affirmation of Freedom of Inquiry and Expression," National Academy of Sciences Resolution, 27 April 1976, quoted in Robert V. Gentry, *Creation's Tiny Mystery* (Knoxville, Tenn.: Earth Science Associates, 1992), 7.

Acknowledgements

For reviewing *The Case Against Darwin* prior to publication, and providing insightful comments and correction, I thank Dr. Wayne Frair, Dr. Don Batten, Dr. Trevor Sadler, Dr. Robert Goette, Mark Stewart, Charles Trotter, W. H. Entz, Barbara Robidoux and Susan Scherer.

I also thank the talented Paul Ingbretson for his continued encouragement and for his contributions to the cover design.

"INTRIGUING"

YOU'VE READ *THE CASE* NOW READ *TORNADO* AND GET *ALL*

Dr. Duane T. Gish, Senior Vice President, Institute for Creation Research: "*Tornado in a Junkyard* by James Perloff should be in the library of every one who is interested in the subject of origins. This book is a powerful argument for creation because it is thorough, fully documented, and scientifically accurate. It is easily readable by scientist and layman alike, and is written in a popular style that will make it interesting and entertaining for readers of all ages. I highly recommend this book."

Massachusetts News: "Perloff's book is a powerful synthesis of recent work by microbiologists, physicists and other scientists showing there is no hard evidence for the creation of new species from existing ones."

Dr. Emmett L. Williams, President, Creation Research Society: "*Tornado in a Junkyard* is a unique presentation of the scientific case against Darwinism, informally written for laymen. If you are looking for a user-friendly explanation of the facts supporting creation, this book is for you."

Conservative Book Club: "James Perloff brings *all* the data together in a volume readily accessible to nonscientific types. . . . Perloff's style, unusually lively, makes *Tornado in a Junkyard* entertaining as well as educational."

Actor Jack Lemmon, who played Clarence Darrow in the 1999 film version of *Inherit the Wind*: "My congratulations to Mr. Perloff for an outstanding piece of work."

The New American: "This is a very important work, written in an informal and attractive style that is a joy to read."

Vicki Brady, Host, "Homeschooling USA": "I recommend that every homeschool family and church have a copy for their libraries."

—*PUBLISHER'S WEEKLY*

AGAINST DARWIN— IN A JUNKYARD
THE FACTS!

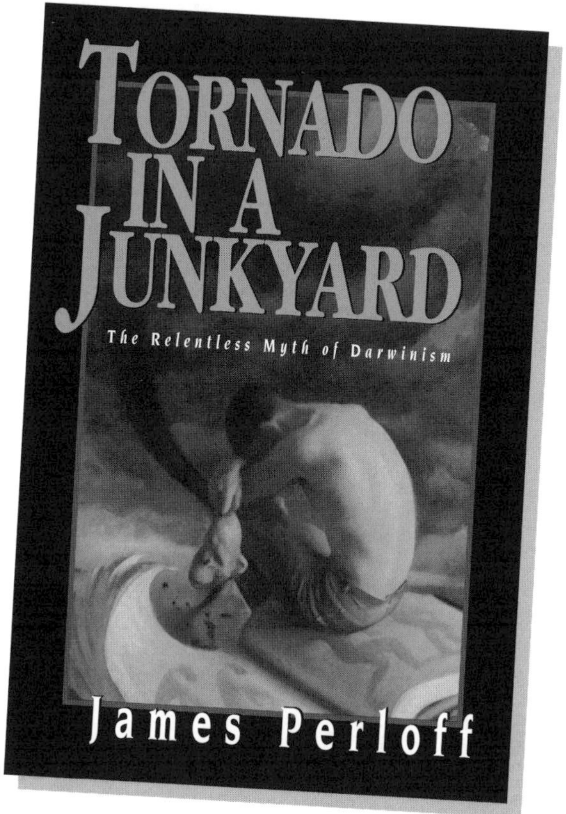

Covers diverse scientific challenges to Darwinism. Compares film script for *Inherit the Wind* to actual court transcript of Scopes Trial. Discusses ape-men in detail—radiometric dating techniques—dinosaurs—much more. 321 pages. Over 80 illustrations. Indexed. $16.95. Paperbound. From Refuge Books.

ORDER FROM YOUR FAVORITE BOOKSELLER